To my mother, Jean McClure Brockenbrough,
for fostering our family's love of science
—M.B.

To Eva, who, like the Saharan dust,
forever changes all she touches
—J.M.-N.

THIS IS A BORZOI BOOK PUBLISHED BY ALFRED A. KNOPF.
Text copyright © 2025 by Martha Brockenbrough
Jacket art and interior illustrations copyright © 2025 by Juana Martinez-Neal

All rights reserved. Published in the United States by Alfred A. Knopf, an imprint of Random House Children's Books, a division of Penguin Random House LLC, 1745 Broadway, New York, NY 10019.
Knopf, Borzoi Books, and the colophon are registered trademarks of Penguin Random House LLC.

Visit us on the Web! rhcbooks.com
Educators and librarians, for a variety of teaching tools, visit us at RHTeachersLibrarians.com

Library of Congress Cataloging-in-Publication Data is available upon request.
ISBN 978-0-593-42842-9 (trade) — ISBN 978-0-593-42843-6 (lib. bdg.) — ISBN 978-0-593-42844-3 (ebook)

Photo (page 37) courtesy of Joshua Stevens/NASA Earth Observatory EPIC Team using NOAA's DSCOVR satellite, June 18, 2020.
The text of this book is set in 17-point Brandon medium.
The illustrations were created using pastels, colored pencils, acrylics, gesso, and fabric on hand-textured paper.

Editor: Rotem Moscovich • Designer: Sarah Hokanson • Copy Editor: Melinda Ackell
Managing Editor: Jake Eldred • Production Manager: Jennifer Moreno

MANUFACTURED IN CHINA   10 9 8 7 6 5 4 3 2 1   First Edition

The authorized representative in the EU for product safety and compliance is Penguin Random House Ireland, Morrison Chambers, 32 Nassau Street, Dublin D02 YH68, Ireland, https://eu-contact.penguin.ie.

Random House Children's Books supports the First Amendment and celebrates the right to read.
Penguin Random House values and supports copyright. Copyright fuels creativity, encourages diverse voices, promotes free speech, and creates a vibrant culture. Thank you for buying an authorized edition of this book and for complying with copyright laws by not reproducing, scanning, or distributing any part of it in any form without permission. You are supporting writers and allowing Penguin Random House to continue to publish books for every reader. Please note that no part of this book may be used or reproduced in any manner for the purpose of training artificial intelligence technologies or systems.

A Gift of Dust

How Saharan Plumes Feed the Planet

Martha Brockenbrough

Illustrated by
Juana Martinez-Neal

Alfred A. Knopf New York

When a sunbeam slips through a window
in a certain slant of light,
you can see a scattered sparkle:
dust!

Once, each speck was something else:
dirt, pollen, or a bit of a living thing
traveling someplace new.
And, depending on where you are,
some of this dust
might have
an even more remarkable story.

Imagine a silvery green trout

gulping plankton

near the bottom

of a giant lake on the African continent.

One day, as all living creatures do,

the fish dies.

Thousands of years pass.

The lake shrinks.

As it dries up,
the trout, and the tiny creatures it once ate,
slowly become fossils.

Wind finds them
in the cracked earth
the lake has left behind.
Steadily, tirelessly, it blows,
and the fossils
of those once-living things
take flight.
They become . . .
dust.

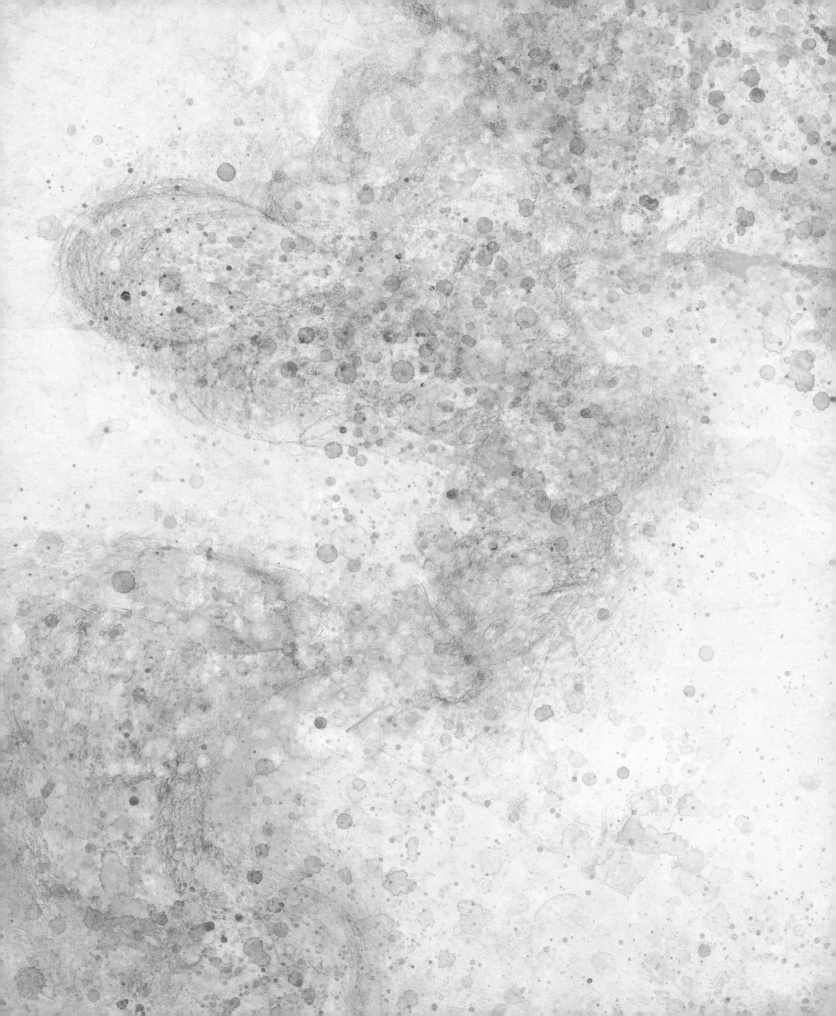

Dust that paints the sky.
Dust that soars across West Africa.
Dust that rides the wind
over the Atlantic Ocean, where . . .

it sifts into the water,

providing nutrients

that help the sea make the air we breathe.

The dust also feeds plankton,

which nourish fish

and sharks

and reefs

and mighty whales.

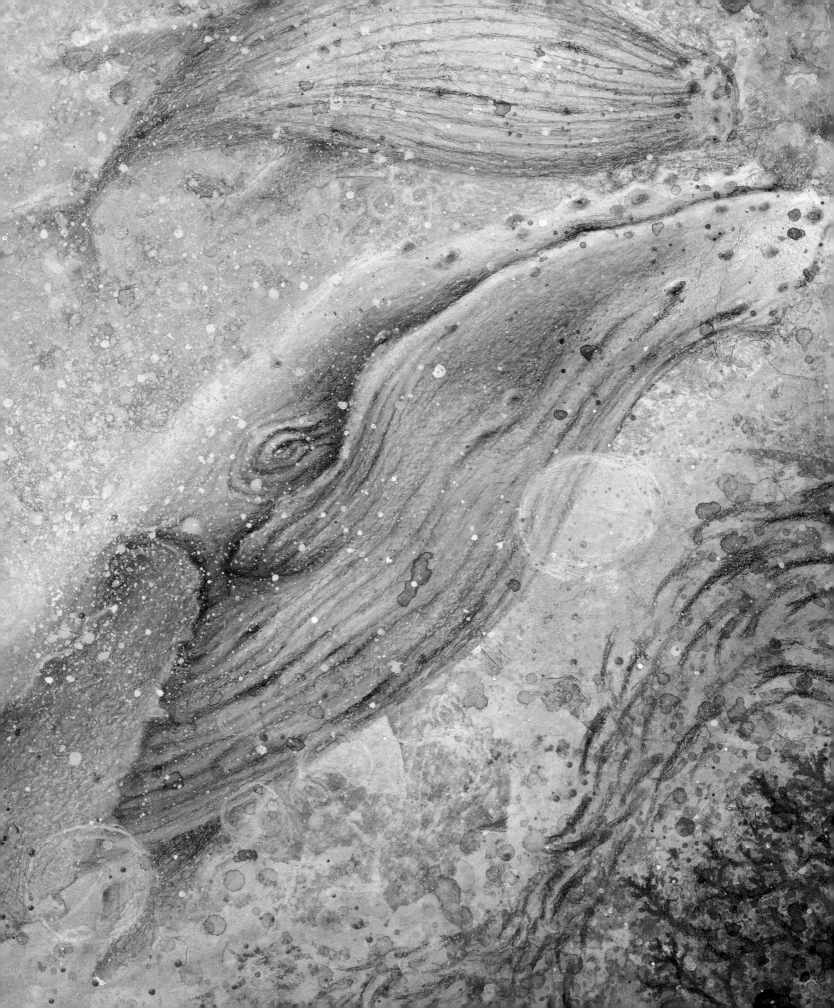

Dust clouds bump into hurricanes.
They soften many storms
on their way
from the driest place on earth
to one of the wettest . . .

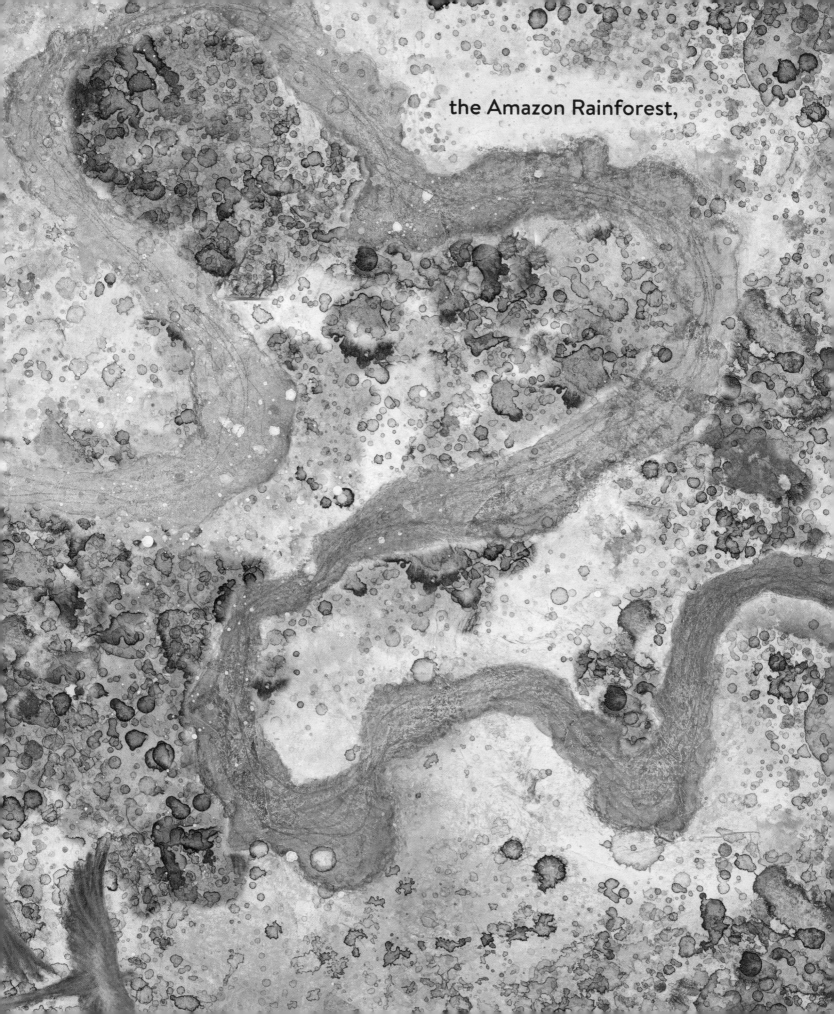
the Amazon Rainforest,

where, in some places, enough rain falls
to submerge a telephone pole,
or a four-story house,
or a stack of twenty capybaras
balancing on each other's backs.

All that rain washes away phosphorus,
a nutrient that plants need
to grow tall and strong.
But the dust brings phosphorus back, making
the rainforest the greenest place on earth.

There's so much dust
that astronauts can spot it from space,
a melon-maracujá swirl,
glittering against the blue-green planet.

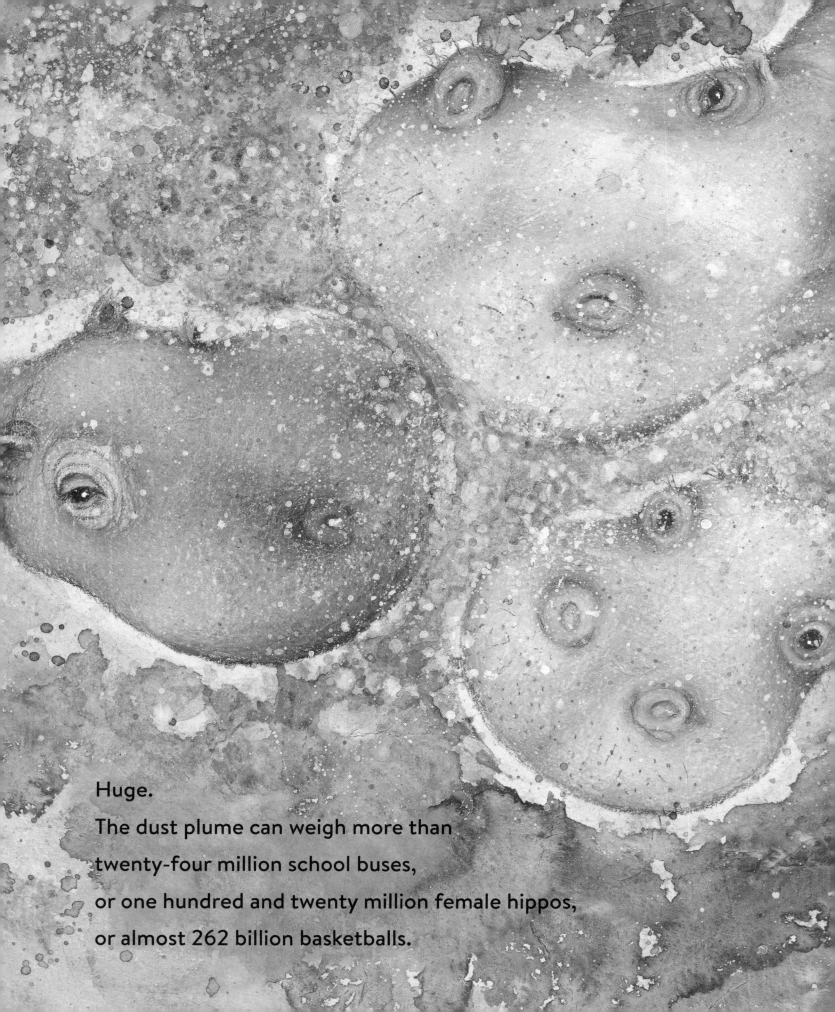

Huge.
The dust plume can weigh more than
twenty-four million school buses,
or one hundred and twenty million female hippos,
or almost 262 billion basketballs.

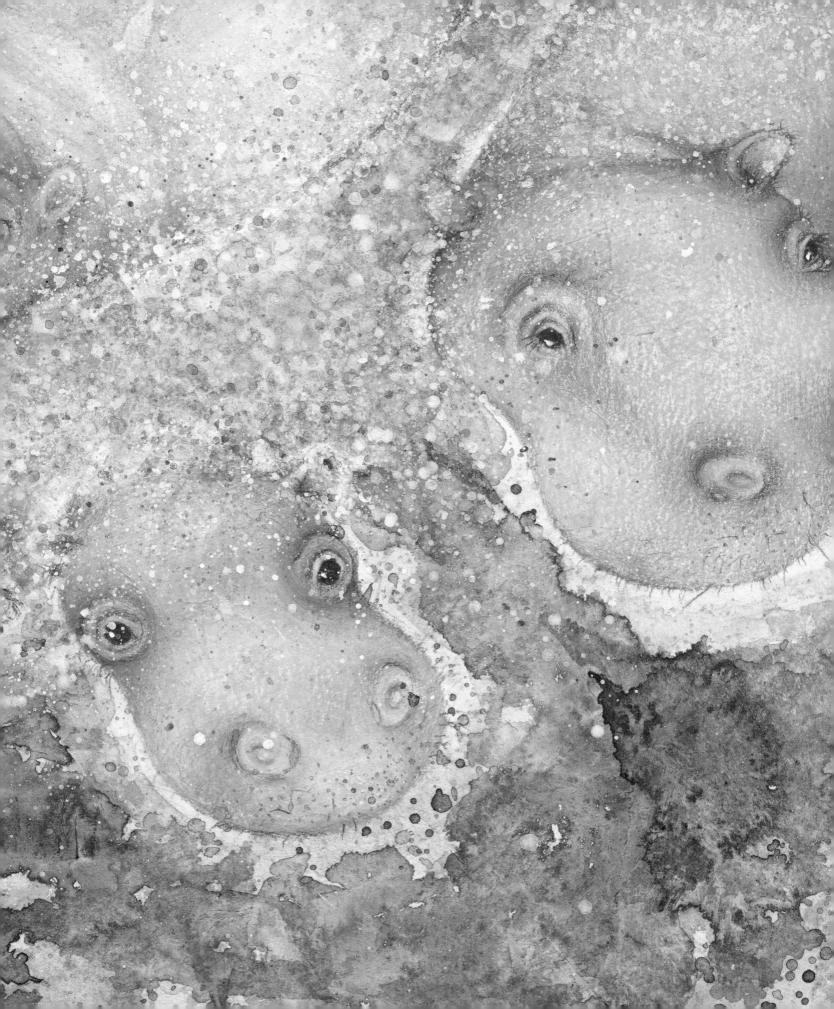

Feeding the ocean and the jungle,
turning sunrises and sunsets red,
it travels to the Caribbean,
the United States,
and beyond.

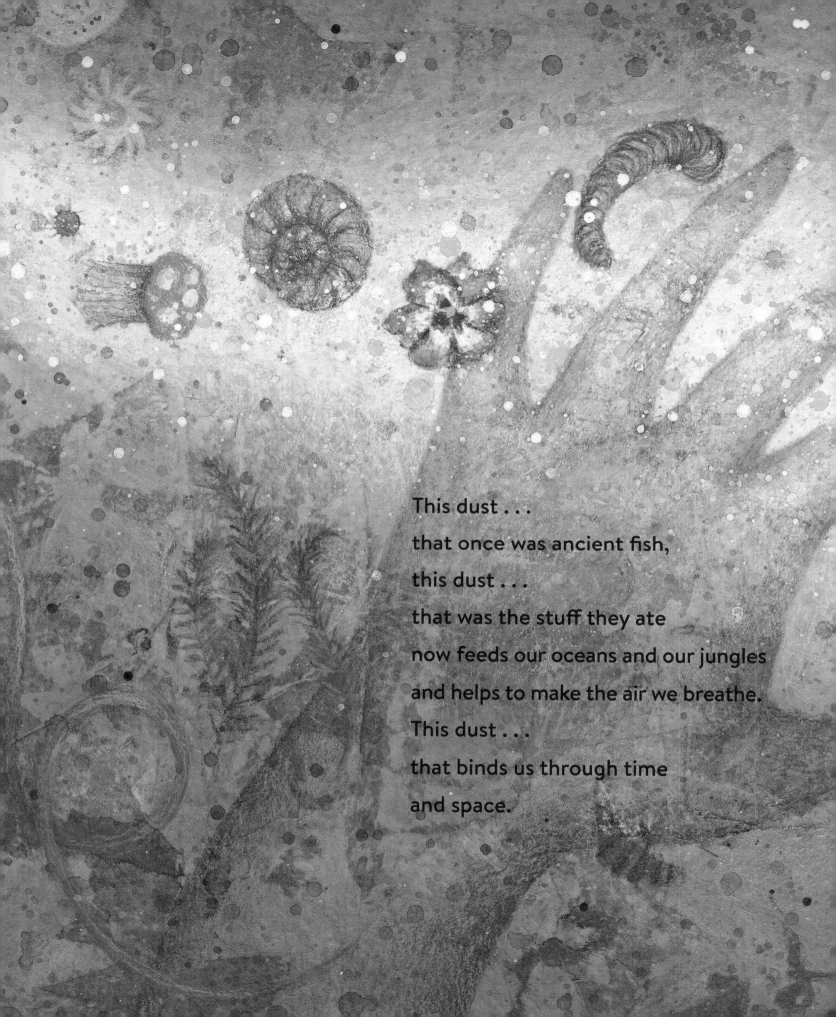

This dust . . .
that once was ancient fish,
this dust . . .
that was the stuff they ate
now feeds our oceans and our jungles
and helps to make the air we breathe.
This dust . . .
that binds us through time
and space.

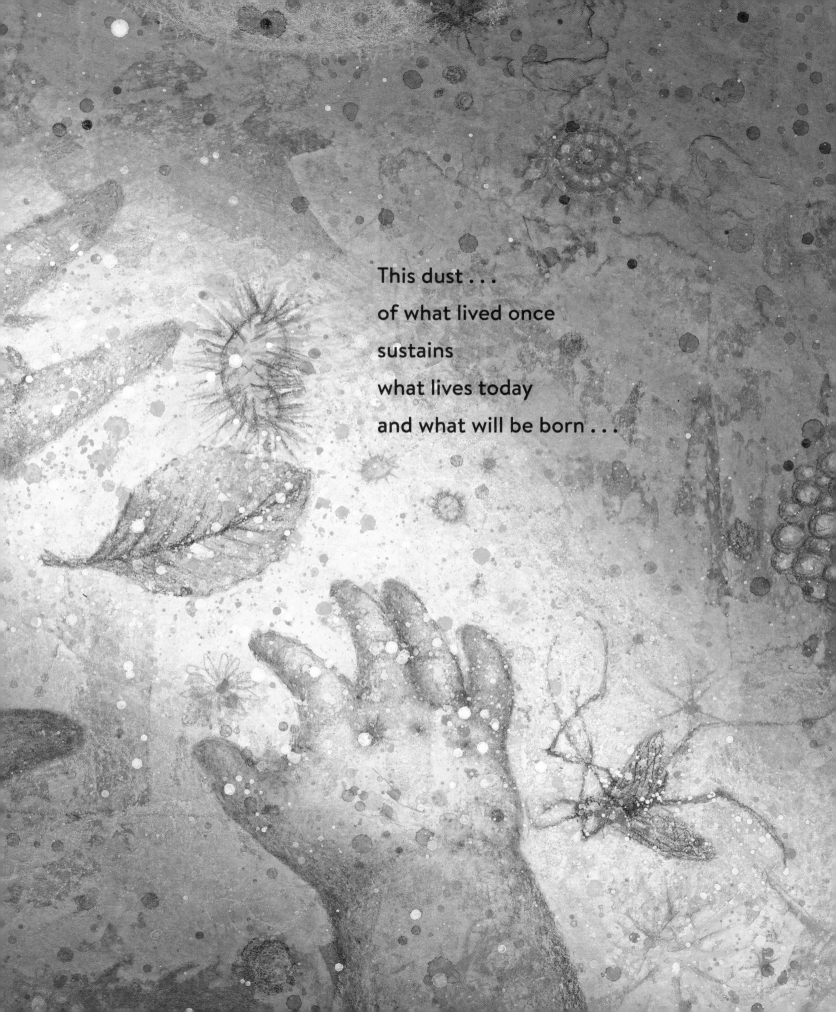

This dust . . .

of what lived once

sustains

what lives today

and what will be born . . .

More Information About Dust

Where is the dried-up lake that makes all the dust?
It's in a North African country called Chad, which got its name from European explorers in the 1800s who didn't realize that "chad" wasn't the name of the place—it was the local Kanuri word for "lake."

The marshy region those explorers visited no longer exists. Nor does what once was an absolutely huge lake. Thousands of years ago, Lake Mega-Chad was as large as Germany. Now it is a paleolake, which means "an ancient lake that has either dried up or shrunk."

Lake Chad is much smaller today—a little bit bigger than Phoenix, Arizona. As recently as sixty years ago, the lake was twenty times that size. It's shrinking rapidly lately because more people live nearby, because it's used for irrigation, and because the earth is warming.

How do fossils form?
Most things that die rot, which is how nature recycles.

Fossils are rare and special reminders of plants, animals, and other organisms that once lived. They can be made of body parts or other traces, such as footprints and even poop. A portion of the dust we're talking about comes from body fossils.

Usually, body fossils form when a plant or animal dies someplace wet and is buried beneath mud or silt. There, the soft tissues—skin and muscle in animals—slowly decay, leaving behind shells or bones.

As more layers of sediment stack up, they press the layers below, turning them into sedimentary rock. As sedimentary rock is squeezed into existence, water seeps into the spaces the animals have left behind. This water leaves minerals that gradually turn to stone, shaped like bones, shells, and teeth.

Fossilization takes a long time: tens of thousands, or even millions, of years.

How does the dust form?
About half the Saharan dust in the atmosphere comes from the Bodélé Depression, a region around Lake Chad. It sits between two mountain ranges that funnel air across the soil. More than a hundred dust storms a year scrape the top layer away, turning it into the dustiest place on earth.

This isn't like the dust in your house, which is most likely animal dander, bug poop, dirt, tiny bits of plastic, and pollen. Your skin also makes dust, but if you bathe, most of that stuff goes down the drain.

How far does the Saharan dust go?
The Saharan dust travels more than five thousand miles: to the Amazon, the Caribbean, Europe, the United States, and East Asia. It's even been found at the North Pole.

How big is the dust cloud and when is it visible?
Dust from North Africa forms a huge band, a mile high and two miles thick. It travels whenever there is wind, and it peaks in spring and summer. The Bodélé Depression produces dust all year round, peaking between January and March and again in late October.

What effect does Saharan dust have on the earth?
Dust from the Sahara worsens air quality and can cause blooms of harmful algae that are called red tides. But it also plays an essential role in nourishing both the Atlantic Ocean and the Amazon Rainforest.

Dust helps make rain. Particles act like little seeds for raindrops, which return the dust to the soil.

Dust from the Sahara contains diatomite, the fossilized remains of single-celled algae called diatoms. Wind scrapes them up and scatters diatomite across the planet. Diatomite does at least five important things:

It puts iron in the ocean. The sea needs this to turn carbon dioxide into oxygen, which we need to breathe. Oceans produce from fifty to eighty percent of all the oxygen in our air.

It helps oceans absorb extra carbon dioxide. Carbon dioxide is a "greenhouse gas," which makes the planet warmer. Too much carbon dioxide in the atmosphere can make the planet dangerously warm. (Dust can also contribute to the warming of the planet, such as when it lands on snow, absorbs sunlight, and makes it melt faster.)

It helps feed phytoplankton, which feed small fish and crustaceans, which in turn feed larger fish, sharks, coral, and baleen whales.

It provides nutrients for the soil in the Amazon Rainforest, which receives an average of nine feet of rain each year—rain that washes away phosphorus that the dust restores. The Amazon is vital for the health of the planet. It produces six percent of the world's oxygen and at least ten percent of the world's plants and animals. Around 380 indigenous populations live there, too.

The dry, dusty air makes some hurricanes less potent, and as climate change makes hurricanes more common, this will reduce their damage.

The Sahara Desert is not the only source of this kind of dust. Someday, it may run out. But somewhere, chances are, another lake will dry out, and the cycle will continue.

Dust traverses the Atlantic Ocean.

For further study:

NASA Earth Observatory: Images from space
> See a collection of images of Saharan dust.
>> earthobservatory.nasa.gov/images/146011/bodele-dust
>
> Read about the Amazon and see photos.
>> earthobservatory.nasa.gov/images/145649/mapping-the-amazon

Books

Desert Biomes Around the World by M. M. Eboch, Capstone Press
> Did you know the Antarctic is also a desert?

Fossil Hunter: How Mary Anning Changed the Science of Prehistoric Life by Cheryl Blackford, Clarion Books
> A biography of one of the world's first paleontologists and fossil hunters

Over and Under the Rainforest by Kate Messner, Chronicle
> A tour of one of the world's most diverse ecosystems

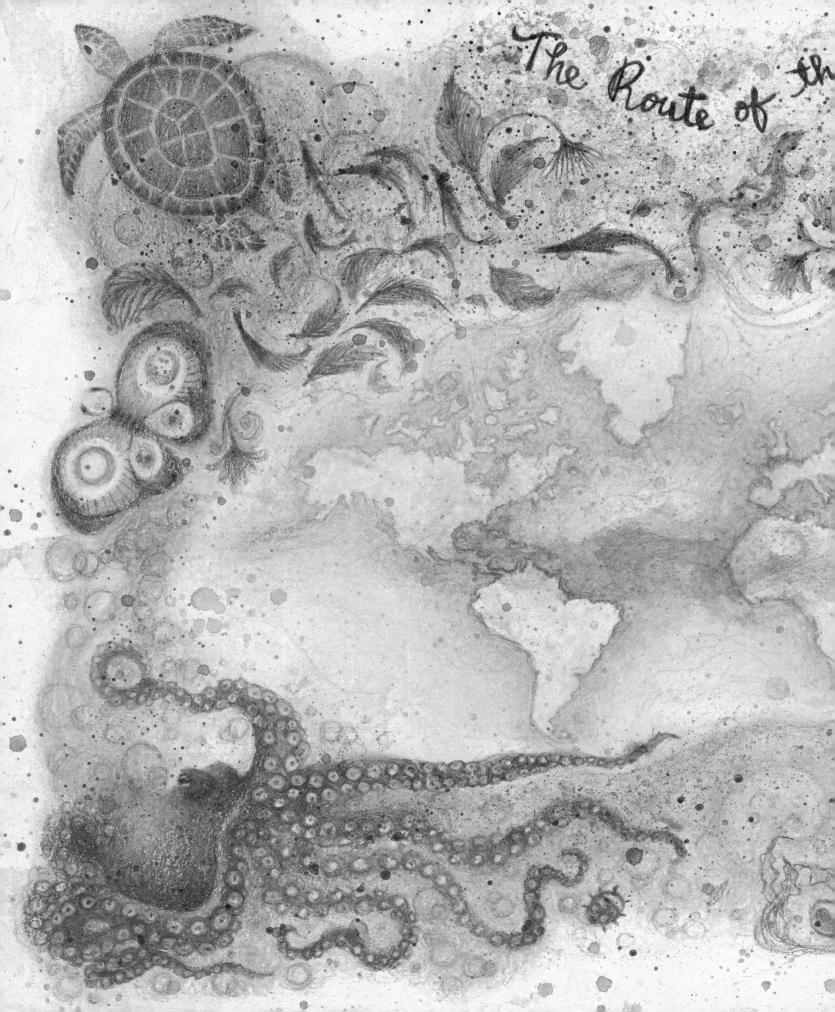